Edge-of-Universe Photo Confirms Quantum-Dimensional Cosmology over "Big Bang"

Lawrence Dawson

The Paradigm Company, Boise Idaho

Content

1. Hubble's "Edge of Universe" Photo Exposes 1960's
   Big-Bang Deception   *Page 1*

2. The Quantum Curvature of Space
   vs. An Expanding Universe
   (comparisons by Hubble's original redshift data)   *Page 12*
   Reprint from the Quantum Dimension

The Snake River N-Radiation Lab
www.srnrl.com

P.O. Box 311 Wilder, ID 83676

Hubble's "Edge of Universe" Photo Exposes 1960's Big-Bang Deception

and Confirms New Quantum-Dimensional Cosmology

Lawrence Dawson
The Snake River N-Radiation Lab

"Quantum-dimensional mathematics are becoming increasingly exact in their predictions and interpretations of physical events, increasingly complex in their mathematical detail, and increasingly alien to the belief systems governing contemporary science."

Lawrence Dawson

"Big Bang" cosmology is built upon a half-century of deception

Hubble's "Edge of Universe" Photo Exposes 1960's Big-Bang Deception and Confirms New Quantum-Dimensional Cosmology

Hubble telescope identifies galaxy at the edge of the visible universe which has two dimensional detail and a visible redshifting of its light.

SOURCE: *"Hubble Sees Infant Galaxies at the Edge of the Universe"* By Phil Plait

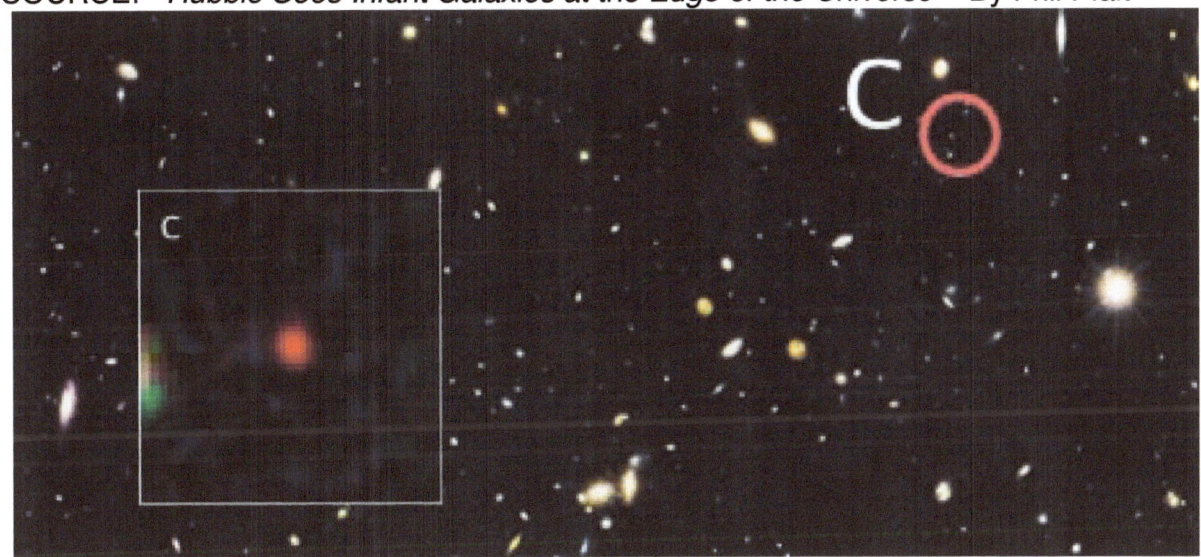

http://www.slate.com/blogs/bad_astronomy/2014/01/09/distant_galaxies_ hubble
_image_of_galaxies_at_the_universe_s_edge.html;

DIGITALLY ENLARGED HUBBLE PHOTO as presented by Plait.

The digital image of the proposed "edge of the universe galaxy" presented by Phil Plait shows the galaxy as having two dimensional characteristics and as being completely red in color.

Although the amount of redshift (i.e. the light's "Z" factor) is not given, the light is completely shifted into the red which is consistent with the quantum dimensional maximum red shifting of "Z=1.5708":
 [maximum quantum-dimensional redshift=(π/2)(wavelength)=1.5708(wavelength)]

Maximum shifting shows that the highest visible "6s" subshell (wavelength=389 nm) shifts to the red/yellow boundary (610.85 nm) and the blue "6g" subshell (wavelength= 486.1 nm) shift to red (763.56 nm) and the red orange "6h" subshell (656.23 nm) shifts out of the visible spectrum into the infrared (1030.8 nm). A "Z" factor of "1.57" would shift 17% of the visible spectrum into the red and the rest into the infrared producing an overall red color which is consistent with what is seen in the Hubble photo.

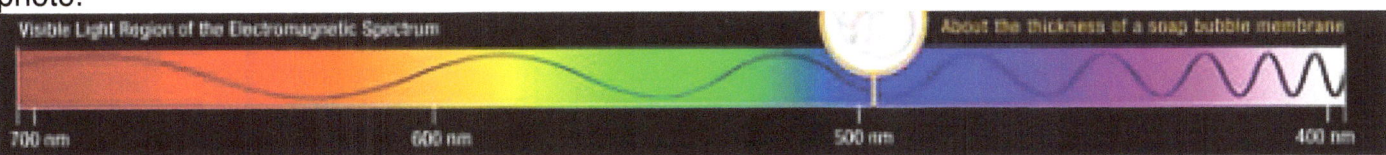

http://missionscience.nasa.gov/ems/09_visiblelight.html

Why Reported Contemporary Stellar Redshift Data is Deceptive

Actually, the measurement of stellar redshift is somewhat simple and straight forward. Most starlight is hydrogen based and the hydrogen absorption lines are a clear indicator of redshift. The wavelengths associated with the visible-light Balmer subshells (the quantum dimensional model)— the "6h" (656 nm) , "6g" (486 nm), "6f " (434 nm) and "6d" (410 nm)— are missing in starlight spectrographs because they have been absorbed. The amount these absorbed wavelengths are shifted upward in wavelength is a measure of redshifting. Despite the utility of the absorption lines from the visible spectrum they have been recently deserted in contemporary astronomy.

> *"In spite of the obvious sensitivity of the Balmer lines to the physical and chemical conditions in the atmospheres of stars, they are underutilized in modern spectrographic analyses. We have been attempting to understand why this is so, and to see what can be done to improve the situation.[1] "*

The use of highly accurate hydrogen absorption lines have been deserted in recent years. For this reason, the amount of redshifting in the Hubble telescope's "edge galaxy" can no longer be credibly analyzed by the largest survey of galactic redshift ever accomplished, that of the Sloan Digital Sky Survey (SDSS).

Redshift should either be accurately measured by shifts in absorption lines or be found to be indeterminate. Instead SDSS has made the amount of redshift "problematic'" which needs to be judged by a computer program. This "computer adjustment" of redshift measurements was identified on the SDSS web site:

> *"[SDSS measuring] methods are empirical in the sense that they use a training set as a reference, then apply machine learning techniques to estimate redshifts. The training sets contain photometric and spectroscopic observations for galaxies. We have chosen to use machine learning techniques with training sets, as opposed to template fitting methods, because of the machine learning techniques' [have] higher overall precision.[2] "* (Italics ours)

That is, the SDSS computer programs are designed to manufacture independent criteria to "judge" the amount of redshifting, rather than being given predetermined *"spectroscopic training sets"* (such as the factual measuring of hydrogen absorption line shifting) which establish a *"template fitting"* for the machine to use.

How can computers self-teach redshift criteria from *"photometric and spectroscopic data"* outside that provided by *"template-fitting spectrographic data";* to judge outside known characteristics such as shifts in hydrogen absorption lines? In what sense does *"machine learning"* provide *"higher overall precision?"*

The answer is instructive. The machine "learns" to fit ambiguous redshift data to the radical and illegitimate revision of *"Big-Bang"* cosmology which occurred during the 1960's.

In 1929, Edwin Hubble had demonstrated that a statistical correlation existed between Cepheid distance measurements to stellar formations and the amount of redshift in the light from those stellar sources. He explained this correlation as Doppler effect from an expanding universe. The "Big Bang" universe was said to be expanding like the surface of a balloon, the points upon which separate or recede from one another at a rate which increases with the distance between any two points. From his data set, Hubble estimated this rate of recession as approximately 485 kilometers per second for each Megaparsec of distance[3].

However, Hubble's expanding universe would also measure the age of the universe and Hubble's recession constant gave too low of an age estimate. After his death, astronomers radically altered the constant to account for an approximate 14 billion year old universe. They reduced his "485 km/sec/Mpc" to approximately "70 km/sec" for each Megaparsec of distance[4] ." This was done without reference to any data (an exo-data revision) and in direct contradiction to Hubble's own data.

---

[1] University of Michigan astronomy web site. *http://dept.astro.lsa.umich.edu/~cowley/balmers.html*

[2] http://www.sdss3.org/dr8/algorithms/photo-z.php

[3] Dawson, L. *"The Quantum Curvature of Space vs. An Expanding Universe; comparisons by Hubble's original redshift data"* p.p. 97-101. *http://www.paradigmphysics.com/*Curvature-Redshift.pdf

[4] Ibid.

The 1960's revision of the expansion constant was strictly a mathematical adjustment such that the recession velocity would not exceed the speed of light at a presumed approximate 14 billion light year maximum separation between two points along the universe's "surface of the balloon." The "$14e\,9$ ly" extent of the universe provided a speed-of-light limit on the expansion constant of "69.84 km/sec/Mpc." Choosing expansion constants which revolved around this value satisfied revisionist, non-empirical astronomy until the arrival of the SDSS which attempted to measure the extent of the universe using the revisionist cosmology with empirical redshift data.

In 2008, after surveying approximately 50,000 galaxies, the survey had found a maximum galactic redshift of "Z=1.4[5]." By applying raw redshift data, SDSS had only identified a distance of 28.6% ($4.0009e\,9$ ly) of the universe's presumed extent, using the '60's exo-data revision's criteria [6]. The survey could not really fulfill its mission of measuring the universe under the assumption of the revised "Big Bang" cosmology and by using actual optical data.

Since 2008, SDSS has replaced the actual optical redshift view of the far universe with a computer generated redshift view. The problem is the following: the 2008 "Z=1.4" maximum optical redshift could identify the edge of a $14e\,9$ ly universe extent if the expansion constant were 20 km/sec for every megaparsec (an actual suggested value on the SDSS web site). However, the universe could not reach an extent of $14e\,9$ ly under such a recession velocity. SDSS's solution was to computer-modify redshift views adjusted to the revised "Big Bang" expansion constant.

SDSS'S "BOSS" COMPUTER-GENERATED REDSHIFT-VIEW *(USING THE '60'S EXO-DATA REVISION)* IS NOT OPTICALLY VISIBLE
"Baryon Oscillation Spectroscopic Survey measures the universe to one-percent accuracy [7]"

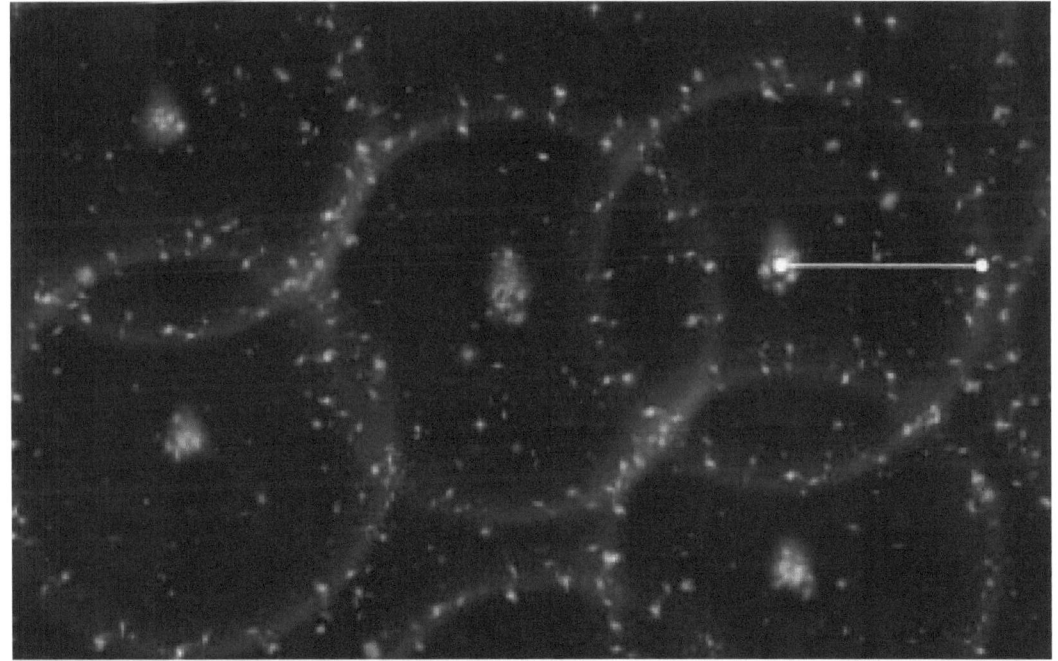

This is an artist's concept of the new measurement of the size of the Universe. The gray spheres show the pattern of the "baryon acoustic oscillations" from the early Universe.

SOURCE: Lawrence Berkeley National Laboratory

---

[5] Forward to *The Quantum Dimension;* L. Dawson; Paradigm Publishing, ISBN 0-941995-24-0, 2009
[6] I had reviewed and reported on SDSS 2008 data in my book *"The Quantum Dimension."* Currently, however, the Sloan Survey has ceased releasing raw redshift data, as they did in 2008, in favor of computer "judged" data.
[7] http://phys.org/news/2014-01-baryon-oscillation-spectroscopic-survey-universe.html

Since 2008, the survey has chosen ambiguous redshift values by a computer program which selects among possible measurements those which are compatible with the 1960's revised Big Bang theory. Doppler-effect redshift proposed by the revised "Big Bang" cosmology would be too great to make "edge galaxies" visible. Therefore, the computer-modified redshift measurements have been focused on closer objects.

The SDSS researchers chose a view of the sky at 2 billion light years distance (by revised Big Bang cosmology). This view allegedly provides data from 12 billion years after singularity rather than the 14 billion years after singularity on the edge of the universe. Theory was concocted which provided that the expansion had undergone variations throughout its history. The 2e9 ly view revealed an expansion variation which had allegedly produced neutrons and protons that were moving near the speed of light and causing light interference. This interference took the form of ring-like undulations in light patterns.

Computers were programed to scan ambiguous redshift data from the targeted distance and to select values consistent with the preprogrammed theory. These selective scans were then used to produce the undulation rings predicted. The radii between the light sources and the edges of the rings, radii which were set by computer data-selection bias, were measured to determine age of the universe. Data which had been biased by theory was then used to prove the theory.

Any high school science student can identify what is wrong with this method. We must assume the staff of the SDSS and the Lawrence Berkeley National Labs have more competence than high school science students. Therefore, we can only conclude this "measurement of the revised Big Bang universe" represents a deliberate deception designed to further support the illegitimate 1960's Big Bang revision.[8]

The Hubble Telescope's Photo of an "Edge Galaxy" Reintroduces Optical Standards to Our Redshift and Distance Analysis

The Hubble telescope's maximum resolution is five hundredths of an arc second or 1.31579e-5° [9]. This resolution simply could not resolve a galaxy of permissible size and redshift characteristics, if the universe were composed by SDSS's revised Big Bang cosmology.

FOR COMPARISON OF SIZE: *Our own Milky Way galaxy is the only galaxy for which we have somewhat reliable estimations of its distance across. We can use this as a standard for comparative purposes. The Milky Way galaxy is 97,800 light years across (approximately 30,000 parsecs):*

$$\{Radius\ of\ Milky\ Way\} \cong 0.0489e6\ \mathrm{ly}$$

Maximum Distance to the Edge of Universe (Quantum vs. SDSS Big Bang)

MAXIMUM QUANTUM ARC DISTANCE AT WHICH LIGHT CAN STILL REACH THE EARTH :[10]
{*light distance to the edge of visible universe*}=$(\pi/2)(48.42\ Mpc)$=247.948943547513e6 ly[11]

SDSS'S PRESUMED MAXIMUM DISTANCE BETWEEN ANY TWO POINTS IN AN EXPANDING UNIVERSE: { max. expansion}=14e 9 *light years*; H_0 = {expansion constant}

$$\langle Calculating\ "H_0"\ for\ "Max. = 14e9\ ly"\rangle:\quad c - \frac{14e9}{3.\,2616e6}H_0 = 0;\quad Mpc = 3.2616e6\ ly$$

---

[8] This is only one example of the many uses of "computer modeling" to force real-world data to fit the pseudo theories advanced by consensus science.

[9] http://hubblesite.org/hubble_discoveries/hstexhibit/telescope/about.shtml

[10] Found by applying the quantum law of elliptical curvature to Hubble's 1929 data table. Space is curved at increasing distance due to increasing tension from the cosmological constant. The Hubble data shows that maximum linear distance for which light curved into an arc can reach the the earth is 48.42 million parsecs. See *"The Quantum Curvature .."* Op .cit.

[11] *"The Quantum Curvature of Space vs. An Expanding Universe; comparisons by Hubble's original redshift data."*   Op.cit

$$H_0 = \frac{c\,(3.2616e6\ ly)}{14e9\ ly} = 69843.1\ \frac{m/\mathrm{sec}}{Mpc}; \qquad H_0 = 69.84\ \frac{km/\mathrm{sec.}}{Mpc}$$

The Potential Optical View of a Galaxy

We know from conventional geometry what the distance of a galaxy must be in order for it to fill the horizon-to-horizon view of our celestial globe. It must be set at the distance of its maximum radius in order for it to fill 180° of our celestial view.

Our View of the Celestial Globe

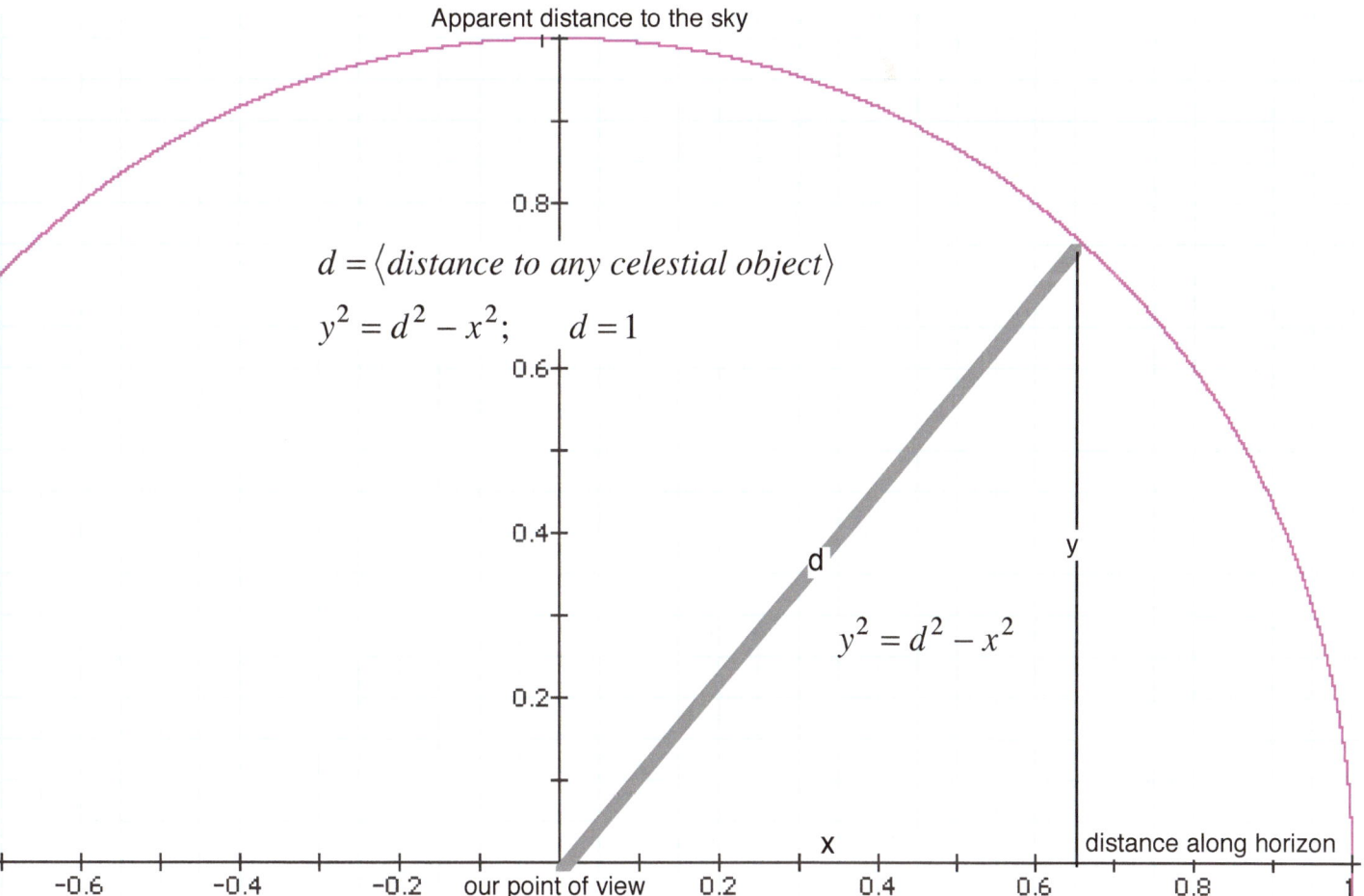

Apparent distance to the sky

$$d = \langle \textit{distance to any celestial object} \rangle$$
$$y^2 = d^2 - x^2; \qquad d = 1$$

$$y^2 = d^2 - x^2$$

BY THE INVERSE SQUARE LAW: The apparent distance to the sky (y^2) from our point-of-view ($x,y=0$) ; this apparent distance changes by the square of the distance along the horizon (x^2). Apparent distance to the sky reaches "0" at "x=d."

For a galaxy set at a distance equal to its maximum radius and with its center at the zenith of our view, the apparent distance to the sky (y^2) along the horizonal axis will reach "0" at the edge of the galaxy. That is, the whole of the galaxy's width will consume the full 180° of our celestial globe when the galaxy is set at a distance equal to its greatest radius.

By the same Inverse Square Law, as the galaxy recedes from us, the angle of our view of the galaxy will resolve as the inverse of the square of the increase in original radial distance. That is, If the distance becomes twice that of the full 180° view of the galaxy, the angle of view will become one-fourth of 180°. The major axis of the galaxy will now only consume 45° of our celestial globe. The mathematical formula for this is the following:

$$r = \{maxium\ radius\ of\ galaxy\}; \qquad d = \{distance\ to\ galaxy\}$$

$$\frac{d}{r} = \{increase\ in\ distance\ as\ a\ multiple\ of\ maximum\ radius\}$$

$$\{angle\ of\ view\} = \frac{1}{(d/r)^2}(180°) = \frac{r^2}{d^2}(180°)$$

Calculating the Radius of an "Edge-Galaxy" as visible to Hubble's Maximum Resolution

d_Q={quantum curvature maximum distance}=247.949e6 *light years*

d_{Ex}={expanding universe maximum distance}=14e10 *light years*; {H_0=69.84}

(Max. Res. Hubble Telescope)=1.31579e-5°

$\{Radius\ of\ Milky\ Way\} \cong 0.0489e6$ ly;

For a Quantum Space curved by a Cosmological Constant[12]

$$\langle For\ Quantum\ Curvature \rangle \qquad \frac{r^2}{d_Q^2}(180°) = \frac{r^2}{(247.949e6)^2}(180°) = 1.31579e\text{-}5°$$

$$r^2 = \frac{1.31579e\text{-}5°}{180°}d_Q^2; \qquad r = \sqrt{7.30994e\text{-}8}(d_Q) = 0.0670e6\ ly$$

$$\{Edge\text{-}Galaxy\ radius\ as\ a\ ratio\ of\ the\ Milky\ Way®\ radius\} = \frac{0.0670e6}{0.0489e6} = 1.37$$

$$\langle Amount\ of\ redshift \rangle \quad Z = \frac{\pi}{2} = 1.5708$$

Quantum Distance for an "Edge-Galaxy" Radius equal to the Milky-Way Radius

$$\frac{r^2}{d_Q^2}(180°) = \frac{(0.0489e6\ ly)^2}{d_Q^2}(180°) = 1.31579e\text{-}5°; \quad d_Q^2 = \frac{(0.0489e6\ ly)^2(180)}{1.31579e\text{-}5°}$$

$$d_Q = 180.864e6\ ly; \quad d_{Q\text{-}max} = \{maximum\ curvature\} = 247.949e6\ ly; \quad d_Q/d_{Q\text{-}max} = 0.72944$$

Calculating the Redshift "Z" Value for Quantum Curvature at 72.944% of Maximum Curvature[13]

d_Q=180.864e 6;   (d_Q)/ ($d_{Q\text{-}max}$)= 0.72944

---

[12] See *"The Quantum Curvature of Space vs. An Expanding Universe; comparisons by Hubble's original redshift data"* @ http://www.paradigmphysics.com/Curvature-Redshift.pdf. As well as *".... the Derivation of Newton's Gravitational Constant from the Quantum-Dimensional Definition of Mass"* http://www.paradigmphysics.com/Quant_Gravity_Model.pdf

[13] See *"The Quantum Curvature of Space vs. An Expanding Universe;......."* p.p. 107-109. Op. cit.

$$d_{lin}(Z) = d_Q; \qquad d_{lin} = \{linear\ distance\}$$

$$d_Q = \{curved\ distance\} = 180.864e6\ ly; \quad Z = 1.41; \quad d_{lin} = 127.895e6\ ly \quad \langle See\ below\ graph \rangle$$

Calculating "Z" and Linear Distance from the Quantum Graph for the Elliptical Curvature of Space

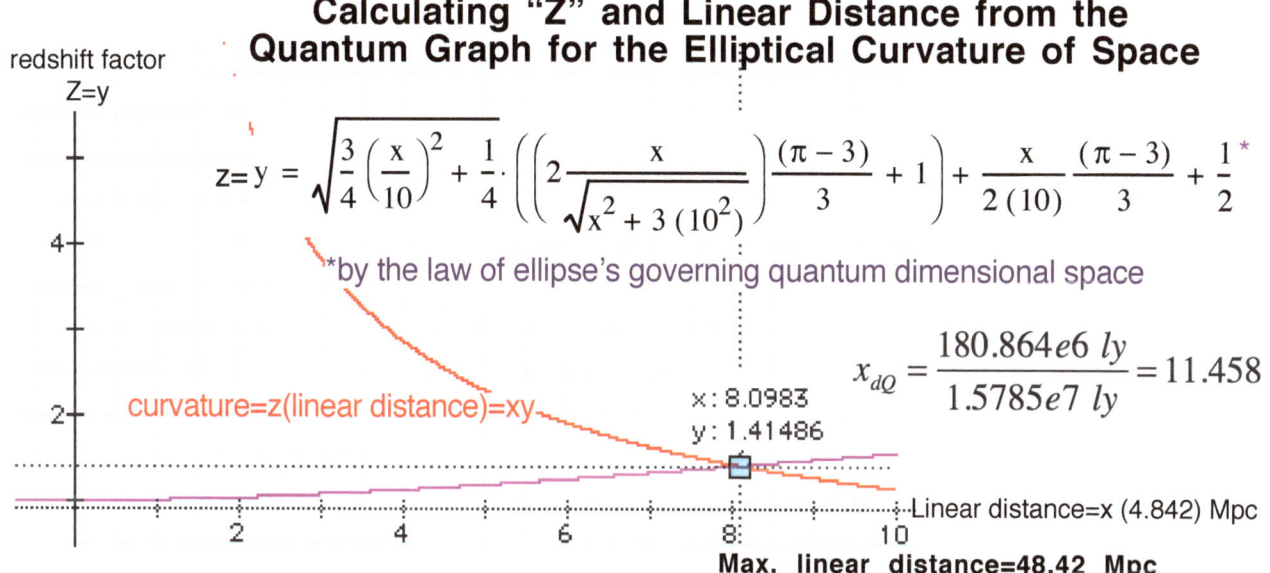

$$z = y = \sqrt{\frac{3}{4}\left(\frac{x}{10}\right)^2 + \frac{1}{4}} \cdot \left(\left(2\frac{x}{\sqrt{x^2 + 3(10^2)}}\right)\frac{(\pi-3)}{3} + 1\right) + \frac{x}{2(10)}\frac{(\pi-3)}{3} + \frac{1}{2}^{*}$$

*by the law of ellipse's governing quantum dimensional space

curvature=z(linear distance)=xy

x : 8.0983
y : 1.41486

$$x_{dQ} = \frac{180.864e6\ ly}{1.5785e7\ ly} = 11.458$$

Linear distance=x (4.842) Mpc

Max. linear distance=48.42 Mpc

Calculating for a Curvature equal to 0.72944 of Maximum Curvature

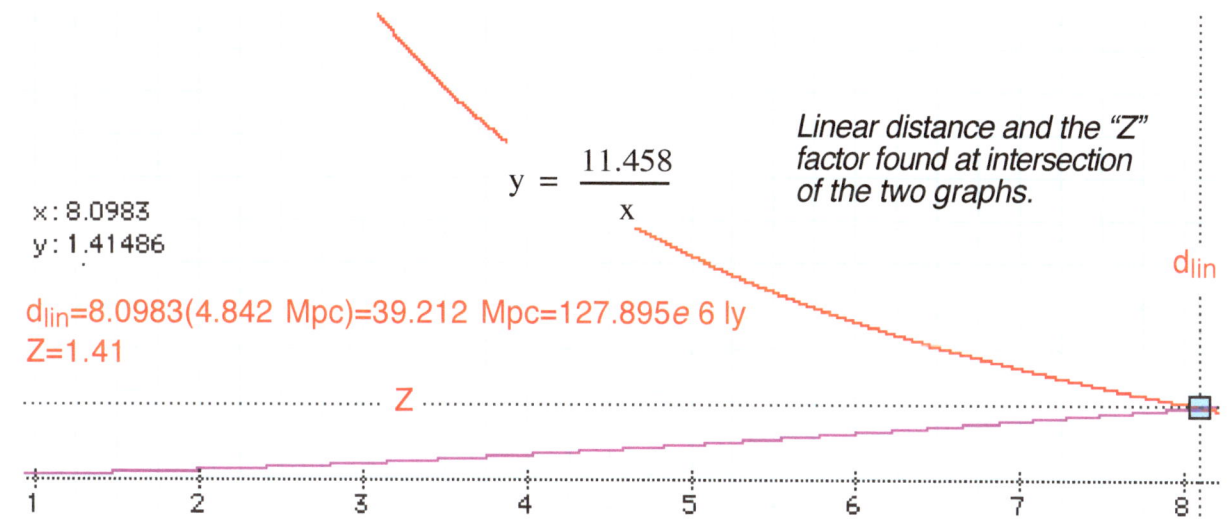

$$y = \frac{11.458}{x}$$

Linear distance and the "Z" factor found at intersection of the two graphs.

x : 8.0983
y : 1.41486

d_{lin}

$d_{lin}=8.0983(4.842\ Mpc)=39.212\ Mpc=127.895e6\ ly$
Z=1.41

Z

For SDSS's Revised "Big Bang" Cosmology

$$\langle Doppler\ Effect\ Redshift \rangle\ Z = \frac{c}{c - H_0(d)}; \quad H_0 = \{Expansion\ constant\} = 69.84\ \frac{km/sec}{Mpc}^{[14]}$$

$$\langle For\ SDSS\ Cosmology \rangle \quad \frac{r^2}{d_{Ex}^2}(180°) = \frac{r^2}{(1.4e10)^2}(180°) = 1.31579e\text{-}5°$$

[14] The Quantum Curvature of Space vs. An Expanding Universe;......." Op. cit.

$$r^2 = \frac{1.31579e\text{-}5°}{180°}\, d^2_{Ex};\qquad r = \sqrt{7.30994e\text{-}8}\,(d_{Ex}) = 3.785166e6\ ly$$

$$\{Edge\text{-}Galaxy\ radius\ as\ a\ ratio\ per\ Milky\ Way\ Radius\} = \frac{3.785166e6}{0.0489e6} = 77.41$$

$$\langle Amount\ of\ redshift\rangle\quad Z \to \infty\quad \langle cannot\ be\ visible\rangle$$

Calculating the Hubble-Resolution Galaxy Radius and % of Maximum Extension for other Redshift "Z" Values

$$H_0^{meters} = 69{,}840\ \frac{m/\sec}{Mpc};\qquad Z = \frac{c}{c - d\left(H_0^{meters}\right)};\qquad Mpc = 3.26163344e6\ ly$$

$$Z(c) - d\left(\left(H_0^{meters}\right)\right)Z = c;\quad d = \left(\frac{(Z-1)c}{Z\left(\left(H_0^{meters}\right)\right)}\ Mpc\right)(3.26163344e6\ ly)$$

$$Let\ "Z" = 1.57;\ d = 5.0831e9\ ly;\qquad \frac{d}{14e9\ ly} = 0.3631$$

$$\{radius\ at\ max.\ Hubble\ resolution\} = r_{HubMax} = \sqrt{7.30994e\text{-}8}\,(d) = 1.374313e6\ ly$$

$$\frac{r_{HubMax}}{r_{MkWy}} = 28.10 \qquad\qquad \langle For\ "Z = 1.40"\ values\ see\ table\ below\rangle$$

Comparing the '60's Exo-Data Revision with Data-Based Quantum Dimensional Curvature for the Hubble Telescopes "Edge Galaxy" at Maximum Resolution

Z Factor	Exo-Data, '60's "Big Bang" Revision $H_0=$ (calculated by '60's standards)				Quantum Curvature Calculated from Hubble's 1929 Data Table			
	distance in light years	% of edge distance	actual λ shift	galaxy radius *per* Milky Way: r=x(Mk Wy rad)	distance in light years	% of edge distance	actual λ shift	galaxy radius *per* Milky Way: r=x(Mk Wy rad)
Max	14e9 ly	100%	infinite	x=77.41	247.949e6	100%	1.57 λ	x=1.37
1.57	5.0831e9	36.31%	1.57 λ	x=28.10	247.949e6	100%	1.57 λ	x=1.37
1.40	4.0002e9	28.57%	1.40 λ	x=22.12	180.846e6	72.9%	1.41 λ	x=1
1.018	247.949e6	1.77%	1.02 λ	x=1.37	20.0964e6	8.11%	1.02 λ	x=0.1111*

Quantum values for "z=y=1.018; x=1.25; curvature =xy=6.161445 Mpc

The '60's Exo-Data "Expanding Universe" Revision— SDSS's Deceptive Computer-Biasing of Optical Data notwithstanding— is Disallowed by the Hubble "Edge Galaxy" Photo

Hubble's Telescope could not see an "edge galaxy" at the maximum extension between any two points in a universe which was defined by the '60's expanding universe revision. The light would not be visible because redshifting would have approached infinity and produced wavelengths which were well outside the range of the Hubble telescope. A dark galaxy at the universe's edge is established by the mathematical formula which produced the '60's revision. Even if such a galaxy were visible, it still would require a questionable radius of 77 *times* the radius of the Milky Way in order to appear with any length and width at Hubble's maximum resolution.

In contrast, the "edge of the visible universe" for the quantum curvature model is the greatest distance for which light can actually reach us. For light sources of greater distance, maximum quantum curvature will cause the light to fall ahead of us along the line of opposition to the distant light source. We will not be able to see it. The quantum curvature is graphed as redshift, "y," to linear distance "x." The graph is built upon Hubble's 1929 data table which showed light redshifting for Cepheids of measurable distances.[15] Hubble's data table was fit to a graphing curve using the discovery of quantum-dimensional mathematics that an exact formula for the periphery of an ellipse existed if the center point of a circle were split to create two focal points with a quantum distance of separation between them[16]. As the the linear distance (major axis of the ellipse) between our view and a light source becomes smaller, the linear distance becomes more "quantum like" with a greater distance of separation between the focal points and a greater eccentricity (smaller variance between elliptical curvature and linear distance). The curvature graph reflects the law of ellipse's for quantum-dimensional space as revealed by the quantum mathematical derivation of an exact formula for the periphery of an ellipse.

However, the quantum-dimensional model of distance related redshifting— as variations in the elliptical curvature of space— cannot be established by either mathematics nor theoretical logic. It must be established by hard empirical evidence and the Hubble telescope's photograph of an "edge galaxy" can provide a key test. The order of magnitude between quantum curvature and the expanding universe has always been immense. Hubble's original expansion constant of around 500 km/sec for every Mpc of distance produced a distance to the edge of approximately 2 billion light years. This seemed too short a time for the age of the universe. This alleged "missing time" initiated the '60's exo-data revision to 14 billion light years. The quantum-dimensional model,' based upon Hubble's data table, predicted a curved distance to the "edge" of 248 million light years. This is a small proportion of both Hubble's original expansion constant and of the '60's exo-data revision (12.4% of Hubble and 1.77% of the exo-data revision). The maximum resolution producing the Hubble telescope's photo of an "edge galaxy" is an excellent chance to test the variances in the order of magnitudes which are predicted by the two models.

Hubble's Resolution of an "Edge Galaxy" confirms the Quantum Curvature Order of Magnitude and disallows the Order of Magnitude proposed by the Exo-Data Revision of an Expanding Universe

The capacity of the Hubble telescope to resolve an "edge galaxy" confirms a visible universe of the order of magnitude proposed by quantum-dimensional mathematics. The telescope has the capacity to resolve a well-defined galaxy of maximum visible redshifting. A galaxy at the visible edge of the quantum-dimensional universe with a radius of 1.37 times that of our Milky Way would be resolved in geometric detail by the Hubble telescope's maximum resolution. Further, the redshifting would be the maximum found in the star field and would shift 17% of the visible spectrum into the red and the rest into the infrared for an overall red which is consistent with what is seen in the Hubble photo. Possible galaxy sizes which ranged between "1.37" times the Milky Way and "1" times the Milky Way incorporate the last 27% of curvature distances, with variations in redshifting between "Z=1.57" and "Z=1.41." At the lower end (galaxy radius equal to the Milky Way's), 27% of the star field would be "redder" than the photo galaxy. At "Z=1.4," 25% of the higher visible frequencies would be shifted between blue and yellow orange while 17% would be shifted into the red and the remainder would be shifted into the invisible infrared frequencies. Unlike the "Z=1.57," the predominant color would not be "red" as found in the "edge"

---

[15] The data table which initiated Hubble's "expanding universe model."

[16] *"Quantum Determination of Elliptical Periphery and the Detection of Systemic Error in the Maclaurin Derivative Series"* L. Dawson, Master's Thesis; http://www.paradigmphysics.com/masters-thesis.pdf

photos. Since 27% of the star field are not "redder" than the "edge" galaxies" the lower 1.4 Z factor must be rejected.

The situation is even more serious for the exo-data revision of an expanding-universe cosmology . The "Z=1.57"— which would allow for the "red" color in the photograph to occur— would position the "edge galaxies" at 36.31% of the 14 billion light year maximum distance between any two points within the expanding universe. 63.7% of total stars should be further away and have a larger "Z" value and be redder in color. Even assuming that other galaxies in the star field do not have radii of 28 + *times* that of the Milky Way which would allow them to be resolved in geometric detail, their light should still be more redshifted than that of the resolved "edge" galaxy. We do not find this in the photo. In fact, the "edge" galaxies were identified as reddish colored light points unique to the star field of which they are part.   The edge galaxies chosen by outstanding red color within the star field, were then resolved to geometric detail by the maximum resolution available to the Hubble telescope. This could not occur if the universe were of the extent and redshift distribution proposed by the 1960's revised "big bang" cosmology as apologized for by SDSS's computer biasing of optical data.

The "Exo-Data" *(outside data)* Interpretations of the Photo and their Absurdity

According to Phil Plait, the author of the article[17], the image is *"part of a survey called GOODS, for Great Observatories Origins Deep Survey, specifically designed to look for the faintest galaxies at the most forbidding distances."* Again, according to Plait, the interpretation given the photo is the following: *"GNDJ-625 [the designation of the galaxy in the photo] is something like 13.2 billion light years away. In other words, we're seeing it when the Universe itself was only about 600 million years old."*

The distance assigned the GNDJ-625 photo is "exo-data" since it could not be determined by redshift of the light. By the exo-data revision of Hubble's Constant to "69.84 km/sec/Mpc," a distance of "13.2e 9 ly" (404.91 Mpc) would produce a "Z" factor of "16.31." "Z=16.31" would lengthen the shortest ultraviolet "root " wavelength (91.143 nm) down to sub infrared at "1458.288 nm." Infrared begins at approximately "820 nm." Since all light emitted by a star has a longer wavelength than the "root" ultraviolet, all light would be shifted to longer than "1458.288 nm." No light would be visible to the Hubble telescope and the GNDJ-625 photo would be impossible.

The "exo-data" radius proposed for the GNDJ-625 galaxy could not possibly be resolved by the Hubble telescope. According to Plait, *"GNDJ-625 itself is only about 1/20th the size of the Milky Way."* [17] However, for the Hubble telescope to geometrically resolve the GNDJ-625 at its maximum resolution of "1.31579e-5°," and, at a distance of "13.2e 9 ly," the radius of GNDJ-625 must be "73 *times*" the radius of the Milky way. Since the Hubble photo clearly resolves the shape of the GNDJ-625, the galaxy cannot be *"1/20th the size of the Milky Way"* using "Big Bang" exo-data assumptions.

The revised exo-data expanding universe, its data-biased support by SDSS, as well as all subsequent "exo-data" interpretations must be rejected as a possible cosmology for the universe.

---

[17]*Hubble Sees Infant Galaxies at the Edge of the Universe,* Plait, Phil.   http://www.slate.com/ blogs/bad_astronomy/2014/01/09/distant_galaxies_hubble_image_of_galaxies_at_the_universe_s_edge.html

The Quantum Curvature of Space vs. An Expanding Universe

Reprinted from the Quantum Dimension

Lawrence Dawson

The Paradigm Company, Boise, Idaho

The Quantum Curvature of Space vs. An Expanding Universe
comparisons by Hubble's original redshift data

The best evidence that linear Euclidean distances in space becomes "kinked" upward to become curved by an intersecting quantum dimension may be the redshift in light frequencies reaching us from far galaxies. The linear distance between us and the source of the light is "kinked" into curvature and light follows the greater curved distance rather than the linear distance producing a redshift "Z". Wavelengths are "stretched" when distances are increased in a single dimension.[127] Light follows the curved arc between "E" and "G" in the strictly Euclidean illustration below.

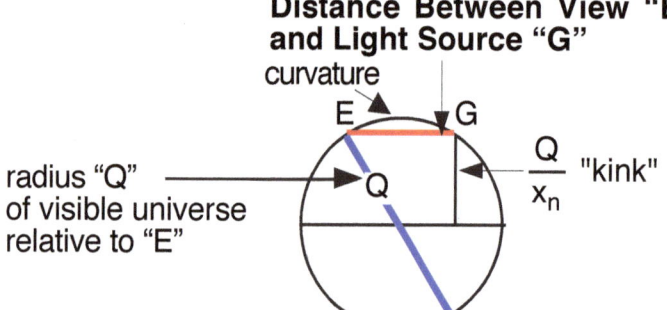

This Euclidean illustration is not an accurate depiction of the quantum curvature of space. The astronomer Edwin Hubble provided the actual mathematical description of curvature, although he never fully understood his contribution. Hubble failed to recognize that the quantum curvature of space gives the universe an appearance of expansion without actually expanding it; that curvature generates an apparent variance in the velocity of light which is the exact equivalent of a recession velocity. Hubble's Constant identifies how the apparent variance in the velocity of light due to curvature varies with distance.

The expanding universe concept is built upon Hubble's discovery that the redshift measured from foreign galaxies is a function of the measured distance to the galaxies and fit a Doppler Effect explanation of the redshift. The Doppler Effect formula[128] for redshift "Z[129]" is the following:

$$Z = \frac{c}{c-v} \quad ; \quad c = \text{speed of light} \ ; \ v = \text{velocity of recession}$$

$$cZ - c = vZ$$

$$v = \left(1 - \frac{1}{Z}\right)c$$

---

[127] This is the principle governing trombones and other wind instruments which are lowered in pitch by lengthening the distance of the air passage.

[128] $f_{rs} = f_o (c - H_o d)/c$; f_{rs} = redshifted frequency ; f_o = original frequency. Taken from the universal formula for the Doppler Effect.

[129] Z=(redshifted wavelength) / (original wavelength)

Hubble's Constant identifies the velocity of recession "v" as a function of the distance to the particular galaxy, hence redshift is also a function of the distance to the galaxy:

Hubble's Constant $= H_o$; d = distance to galaxy in mega parsecs (Mpc).

$$v = H_o d$$

$$Z = \frac{c}{c - H_o d}$$

Hubble didn't realize that the apparent change in velocity for the speed of light "c" due to the forced curvature of a linear distance is nearly the same as the Doppler equation for recession velocity, especially for the close distances of his data table. For the curvature hypothesis, redshift "Z=(curvature distance)/(linear distance)." Both curvature distance and linear distance are measured between the observation point and the light-source:

Redshift by Quantum Curvature and its "Apparent Velocity of Recession"

$d = \{linear\ distance\ to\ light\ source\}$; $\chi = \{curved\ distance\ to\ which\ "d"\ is\ kinked\}$; $Z = \frac{\chi}{d}$; $\{Redshift\} = Z(\lambda)$

$t_1 = \{time\ across\ linear\ "d"\ at\ "c"\}$; $t_2 = \{time\ across\ curved\ "\chi"\ at\ "c"\}$; $\Delta v = \{apparent\ recession\ velocity\}$

$d/t_1 = c$; $\quad \chi/t_2 = c$; $\quad d/t_1 - \Delta v = d/t_2$; $\quad \Delta v = d/t_1 - d/t_2$;

$$\frac{d}{t_1} = c = \frac{9.715612e\text{-}15\ Mpc}{sec.}; \quad t_1 = \frac{d}{9.715612e\text{-}15\ Mpc}; \quad \frac{\chi}{t_2} = c = \frac{9.715612e\text{-}15\ Mpc}{sec.}; \quad t_2 = \frac{\chi}{9.715612e\text{-}15\ Mpc}$$

$$\Delta v = \frac{d}{d/9.715612e\text{-}15\ Mpc} - \frac{d}{\chi/9.715612e\text{-}15\ Mpc} = \left(1 - \frac{d}{\chi}\right)9.715612e\text{-}15\ Mpc = \left(1 - \frac{1}{Z}\right)c$$

That is, the apparent change in velocity " Δv " for the speed of light in a static, curved-space universe might be mistaken for Doppler recession velocity "v" in an expanding universe.

It is generally not recognized that Hubble's Constant —as used to convert redshift to the distance to the light source — is actually a time value (*distance/velocity=time*[130] ; as above). Hubble's Constant is measured in "velocity *per* unit of distance." Specifically, it is velocity in "kilometers *per* second (km/ sec)" *per* distance in "megaparsecs (Mpc[131])":

d = distance ; v = velocity ; t = time

$$H_o = \frac{v}{d} \quad ; \quad \frac{d}{v} = t \quad ; \quad \frac{v}{d} = \frac{1}{t} \quad ; \quad H_o = \frac{1}{t}$$

$$v = H_o d = d/t$$

Hubble's Constant is generally not thought of as a time value because its units of measure are different for "velocity (kilometers *per* second)" and "distance (megaparsecs)." Its application is thought restricted to "recession velocity" for an expanding universe as so many "kilometers/ second" *per* "megaparsec." Distance can be easily found by converting redshift to recession velocity (by Doppler formula) and dividing it by Hubble's Constant. The different units of measure are irrelevant to this conversion:

$$d = \frac{c}{H_0}\left(1 - \frac{1}{Z}\right) \quad \{see\ formula\ on\ opposite\ page\}$$

The fact that Hubble's Constant is a time value becomes extremely significant because it supplies a time period during which the universe has been expanding and, therefore, an alleged time since the hypothesized "Big Bang."

---

[130] A common formula for velocity, distance and time.

[131] 1 Mpc=3 .261 63626 10^6 light years =3.08568025 10^{19} kilometers.

Convert Hubble's Constant from "kilometers *per* second" to "Mpc per second":

$$H_O = \frac{500\,km/sec}{Mpc} = \frac{1.6204e\text{-}17\,Mpc/sec}{Mpc} = \frac{1.6204e\text{-}17}{sec} = \frac{1}{t} \quad \text{[132]}$$

$$t = (6.1713157245e16 \text{ seconds}) = (1.955572e9 \text{ years})$$

$$\{\text{Time of the universe's expansion is approximately 2 billion years}\}$$

$$\{recession\ velocity\} = H_o d = \frac{d(in\ Mpc)}{(6.1713157245e16 \text{ seconds})}$$

$$\left\{ \begin{array}{l} \text{Velocity equals the distance of the stellar light source from the Earth} \\ \text{divided by the amount of time the universe has been expanding.} \end{array} \right\}$$

The "Apparent" Recession Velocity of the Curved Space Model

Recession velocity of Hubble's "Big Bang" cosmology is the distance to the foreign stellar object divided by the time since the beginning of the universe's expansion. The "apparent recession velocity" of the static, curved space model is the change in velocity calculated by light traveling the path of "kinked curvature" rather than the direct path. That change in velocity is equal to the negation of subdivision for the speed of light with the negation factor being the "Z" factor. This is exactly the same formula as for Doppler redshifting velocity.

$$t_1 = \{time\ across\ linear\ "d"\ at\ "c"\}; \quad t_2 = \{time\ across\ curved\ "\chi"\ at\ "c"\}; \quad \Delta v = \{apparent\ recession\ velocity\}$$

$$d/t_1 = c; \quad \chi/t_2 = c; \quad d/t_1 - \Delta v = d/t_2; \quad \Delta v = d/t_1 - d/t_2 = \left(1 - \frac{d}{\chi}\right) 9.715612e\text{-}15\ Mpc = \left(1 - \frac{1}{Z}\right)c$$

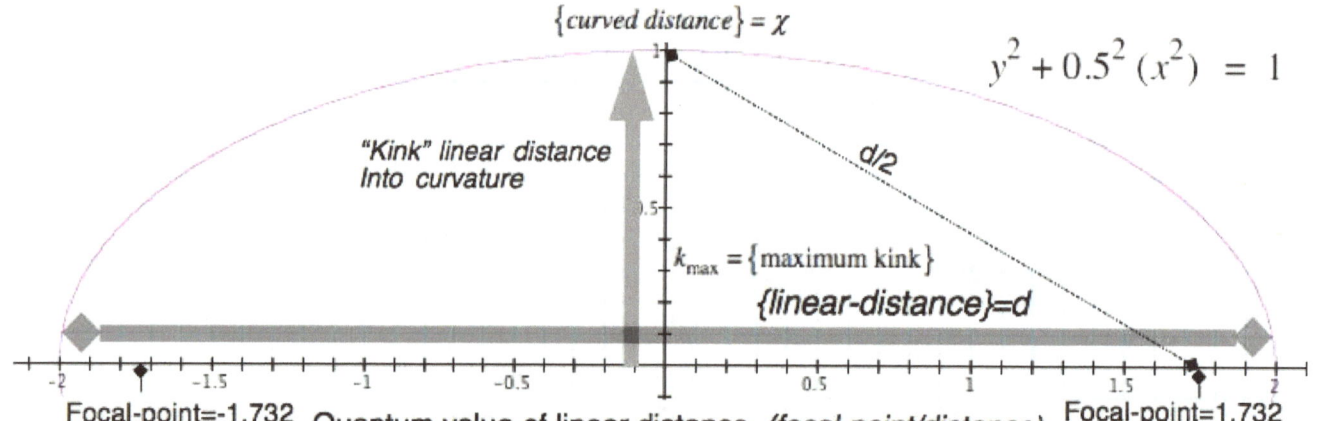

$$\{curved\ distance\} = \chi$$

$$y^2 + 0.5^2(x^2) = 1$$

"Kink" linear distance Into curvature

d/2

$$k_{max} = \{maximum\ kink\}$$

{linear-distance}=d

Focal-point=-1.732   Quantum-value of linear distance=(focal-point/distance)   Focal-point=1.732

The Quantum Law of the Elliptical Curvature of Space

The distance, "d" between any two objects in space composes a line which must be "kinked" into curvature by quantum force. As distance increases, the line becomes less quantum and more Euclidean, increasing the amount of "kinked" curvature to the limit of "(π/2)d" at a maximum cosmological quantum distance (current estimated value 48.42 Mpc). The quantum value of any distance composes the focal points of an ellipse as "Q/d=2(focal point)/d." The distance is "kinked" into elliptical curvature with an eccentricity value which is equal to the negation of subdivision (squared) for the ratio of the distance to the maximum cosmological quantum.

[132] 1 km = 3.24077929e-20 Mpc

14

The Quantum Limit on Visibility across the Universe

The curvature of space restricts the distance an object may be seen from any single viewpoint. All cosmological distances are "kinked" into curvature and the maximum ratio of curvature to linear distance ("Z") is "$\pi/2=1.5708$." Greater distances cannot be kinked into greater curvature. This establishes the limit at which an object is visible and provides the maximum macro quantum value.

$$\varepsilon^2 = \left(1 - \frac{d^2}{Q_{Max}^2}\right); \qquad \varepsilon = \{eccentricity\}; \qquad\qquad Q_{Max} = \{Maximum\ Cosmological\ Quantum\}$$

$$\frac{d}{Q_{Max}} = \{ratio\ of\ distance\ to\ Maximum\ Cosmological\ Quantum\}; \qquad \phi = \{elliptical\ focal\text{-}point\}$$

$$\varepsilon = \{elliptical\ eccentricity\} = \{quantum\ value\ of\ "d"\} = \frac{2(\phi)}{d}$$

$$\phi^2 = \frac{d^2}{4} - k_{Max}^2; \qquad \varepsilon^2 = \left(1 - \frac{d^2}{Q_{Max}^2}\right) = \frac{4\phi^2}{d^2} = \frac{4\left[(d^2/4) - k_{Max}^2\right]}{d^2} = 1 - \frac{4k_{Max}^2}{d^2}$$

$$\left(1 - \frac{d^2}{Q_{Max}^2}\right) = \left(1 - \frac{4k_{Max}^2}{d^2}\right); \qquad \frac{4k_{Max}^2}{d^2} = \frac{d^2}{Q_{Max}^2}; \qquad\qquad \frac{k_{Max}^2}{d^2} = \frac{d^2}{4Q_{Max}^2} \quad [133]$$

$$2\chi = \{circumference\ of\ ellipse\}; \quad r_1 = \{minor\ axis\} = k_{Max}; \quad r_2 = \{major\ axis\} = d/2$$

$$\chi = \sqrt{3k_{Max}^2 + (d/2)^2}\left(\frac{2k_{Max}}{\sqrt{k_{Max}^2 + 3(d/2)^2}}\left(\frac{\pi-3}{3}\right)+1\right) + \left[k_{Max}\left(\frac{\pi-3}{3}\right)+d/2\right] \quad [134]$$

With a Known Maximum Cosmological Quantum, Redshift "Z" is a Function of Distance "d"

$$Z = \frac{\chi}{d} = \sqrt{\frac{3}{4}\left(\frac{d}{Q_{Max}}\right)+\frac{1}{4}}\left(\frac{2d}{\sqrt{d^2 + 3Q_{Max}}}\left(\frac{\pi-3}{3}\right)+1\right)+\left(\frac{d}{2Q_{Max}}\left(\frac{\pi-3}{3}\right)+\frac{1}{2}\right)$$

Hubble's Constant is Revised, without Data, to give a Longer Age of the Universe

Since Hubble's death in 1953, the "Constant" has been continuously adjusted downward from Hubble's empirically determined "500 km/ s/ Mpc" to the current estimate of "65-50 km/ s/ Mpc.[135]"

In 1956, Allan Sandage, Hubble's successor at the Mt. Wilson and Palomar Observatories, began the revisions downward. Sandage revised Hubble's "500 kg/ s/ Mpc" to "180 km/ s/ Mpc." In 1958 Sandage published a value of "75 km/s/Mpc," and by the early 1970's estimates from Sandage and his longtime collaborator Gustav Tammann were hovering around "55 km/s/Mpc[136]," or very near the modern accepted range.

In the National Institute of Standards and Technology report "CODATA Recommended Values of

---

[133] The maximum "kink" is a direct function of source distance and visible maximum quantum across the universe.
[134] "The Quantum Formula for the Perifery of an Ellipse;" Dawson,L http://paradigmphysics.com/masters-thesis.pdf
[135] Harvard University web page "Hubble's Constant."
[136] ibid.

the Fundamental Physical Constants: 2006"[137] the authors give two reasons for the revision of scientific constants besides improvement in measurement and lab techniques. One of these is "time variation of the constant[138]" However, they admit that "there has been *no laboratory observation of time dependence* of any constant that might be relevant to the recommended values" (italics mine). The Hubble revisions are "time variation" revisions without "laboratory observation."

Light redshift due to quantum curvature produces an "apparent recession velocity" because of the variance in time across the curved pathway relative to time across the linear pathway.This "apparent recession velocity" is the equivalent of Hubble's recession velocity in that it resolves to the same formula as the negation of subdivision of the speed of light with the negation factor being red-shifting "Z." The "Z" factors are not equivalent due to the time variations for expanding vs curvature theory.

$$\{recession\ velocity\} = \left(1 - \frac{1}{Z_{Hubble}}\right)c \neq \{apparent\ recession\ vel.\} = \left(1 - \frac{1}{Z_{curvature}}\right)c$$

$$Z_{Hubble} \neq Z_{curvature}; \quad 1/t_{actual\ rec.} = H_O; \quad 1/t_{appear\ rec.} = (t_{curve} - t_{linear})/(t_{curve})(t_{linear}); \quad v_{rec.} = d/t$$

Both the static, quantum curvatured universe and the expanding, Doppler Effect universe provide the same redshift mathematics. The great divide between the two hypothesis —and the only effective test of them— is in their distinct treatments of time.

The hypothesis that the volume of vacuous space is determined by the quantum squared —with redshift being explained by the forced curvature of space— replaces Edwin Hubble's expanding universe concept. Although the quantum curvature model supplies an "apparent recession velocity," it can be seen that the curvature model will predict a different "Z" value than Hubble's model when applied to the Doppler effect formula. This is so because the two models will supply different time values for the same distance. The time value of Hubble's recession velocity is a constant while curvature time is a variable.

Redshift can be explained as Doppler Effect for both. Using the standard Doppler Effect formula[139], the difference between the Hubble and curvature "Z" values is the following:

$$Z = \frac{c}{(c - v_{rec})}; \quad Z_{Hubble} = \frac{c}{\left(c - \dfrac{d}{t_{expansion}}\right)}; \quad Z_{curvature} = \frac{c}{\left(c - \dfrac{d}{(t_{curve})(t_{linear})/(t_{curve} - t_{linear})}\right)}$$

Hubble's "Z" is a direct function of distance "d" because recession velocity time is a constant. It is the time of the universe's expansion. For the curvature model, "apparent recession time" is a function of both "d" and the increase in curvature as "d" increases. At some point, quantum "Z" will increase faster than Hubble's at equivalent distances. Even so, Hubble's original distance-exaggerating time constant was later thought to be too low for the assumed age of the universe[140].

The Proposed Geometry of Hubble's "Big Bang" Cosmology

Hubble's "Big Bang universe" requires expansion over time in order for the universe to have reached its current size. As the distance to any galaxy "G" increased relative to us its velocity of recession also increases. Current velocity is determined by the time of expansion.

---

[137] National Institute of Standards and Technology, Gaithersburg, Maryland 20899-8420, USA; report authors: Peter J. Mohr , Barry N. Taylor , and David B. Newell:
http://physics.nist.gov/cuu/Constants/codata.pdf

[138] *Ibid.* Page 5

[139] $f_{rs} = f_o (c - H_o d)/c$; f_{rs} = redshifted frequency ; f_o = original frequency.

[140] Harvard web site. *op. cit.*

Hubble's "Big Bang" as Curvature Around an Expanding "4-D" Radius

$$r = \frac{\Delta r}{sec.}t \quad ; t = \text{time of expansion} \quad d = \frac{\theta}{2\pi}2\pi r = \theta(r) = \theta\left(\frac{\Delta r}{sec.}t\right); \quad \frac{d}{t} = \frac{\theta \Delta r}{sec.}$$

$$\frac{d}{t} = \{\text{recession velocity}\} = H_o(d) = \frac{\theta \Delta r}{sec.} \quad \langle\text{See pg. 96}\rangle$$

Space said to curve back upon itself.

Fourth dimensional radius, "r," around which universe is said to be expanded

Distances between stellar formations and our view represent expansions from the singularity point over the time which the universe has been expanding. Time is "1/(Hubble's Constant)."

Radial Expansion Velocity is Calculable from Model

$$\langle\text{Maximum expansion is at "}\theta = \pi.\text{"}\rangle \quad \frac{d}{t} = c = \frac{\pi(\Delta r)}{sec.}; \quad \frac{\Delta r}{sec.} = \frac{c}{\pi} = 9.5426903e7 \; meters / sec.$$

Distance "d" is the Integral of Expansion Velocity over the Time of Expansion

$$D(d) = \frac{\theta \Delta r}{sec.}; \quad d = \int_0^t \frac{\theta \Delta r}{sec.}d(t) \quad t = \frac{1}{(Hubble's \; Constant)} = \frac{1}{H_o}$$

Any two points in Hubble's universe have a fixed angle of curvature "θ." This angle remains constant as the distance between the two points expands. The velocity of recession between the two points also has a fixed rate because the expansion of the 4-D axis "r" is at a fixed velocity. This angle *times* fourth dimensional radial expansion velocity *times* the time since the beginning of the universe *equals* the distance factor for the recession velocity. That is, current distance "d" is the integral of radial velocity *times* the angle as integrated over the time since expansion commenced.

"Hubble's Constant" is not actually a *"constant."* It is a *"variable."* It is the time factor since the "Big Bang" and, as such, it is only a *"contemporary value."* That is, Hubble's "Constant" will change as time progresses.

As the time since the alleged "Big Bang" increases, Hubble's "Constant" will get smaller because the constant is actually equal to "1/t." As time increases, the constant will get smaller. Hubble's "Constant" was actually an empirically determined estimate of the time since the alleged "Big Bang" as based upon measured stellar distances and the measured redshifting of their light. Hubble had used a recently discovered method of determining stellar distances greater than those which could be determined by standard parallaxes. Parallaxes uses the visual angles to nearby stars from opposite sides of the earth's orbit to triangulate distances. However, the limit on parallaxes is "0.0001 Mpc" which is much too close to acquire valuable redshift data.

Hubble's Expanding Universe Hypothesis and his Empirical Estimation of the Time of Expansion using Stellar Distance and Redshift Data

In the early 1900s, the Harvard Observatory astronomer, Henrietta Leavitt, had discovered that the pulsation periods of variable Cepheid stars were directly related to their absolute luminosity. Using the Leavitt Cepheid brightness relationship, Hubble was able to measure distances up to 2 Mpc which were "20,000 *times*" greater than parallax maximum distances.

In a study of periodic variables in both the Large Magellanic Cloud as well as the Small Magellanic Cloud, Henrietta Leavitt had classified "47 " Cepheids of measurable periodicity[140B]. From this data, she recognized that the brightness of the Cepheid was related to its periodicity. The longer the period of the Cepheid pulse, the greater the luminosity. From this discovery, Leavitt had developed her periodicity to luminosity table.

It was to others that the realization that Leavitt's discovery could measure stellar distance would fall. The Danish astronomer, Ejnar Hertzsprung, first used Leavitt's data to calculate the distance to Cepheids in the Small Magellanic Cloud (SMC) using the Law of the Inverse Square to compare Leavitt's absolute luminosity with the apparent luminosity from our view.

The American, Harlow Shapley statistically recalibrate the Cepheid absolute magnitude scale using 230 Pulsating Cepheids located in globular star clusters ("*extremely remote and highly concentrated stellar systems, arranged in a spherical form and consisting of tens of thousands of stars"*).[140C] Shapley's large number of globular cluster Cepheids revised the absolute luminosity scale and modified Hertzsprung's distance calculations.

Although Shapley's globular cluster data is best known for correctly identifying the shape and magnitude of our Milky Way galaxy, his data also extended the stellar distances to which absolute magnitude could be applied. He used his more accurate Cepheid luminosity calculations to identify the absolute luminosity of the brightest stars of globular clusters which contained Cepheids of known luminosity. This extended the range of known luminosity to global clusters which were too distant to view detectable Cepheid periodicity. By using globular luminosities, Shapley was able to identify the actual scale of the Milky Way galaxy for the first time (currrently about 30 Kpc or 100,000 light years).

In 1924, Hubble was using the absolute luminosity scales to measure distances of "0.276 Mpc" or many times the distances measured by Shapley in determining the size of our galaxy. By 1929, Hubble was able to push "absolute luminosity" stellar distance measures to "2 Mpc" or 7.25 *times* his 1924 measurements. This allowed him to construct a 24 entry data table of Cepheid absolute luminosity distances which ranged from the SMC's "0.032 Mpc" to the "2 Mpc" of the galaxies "4382, 4472, 4486 and 4649." These distances gave a testable range against measured redshift of light which could then be compared with distance.

Hubble's data comparing redshift with distances as determined by absolute luminosity did not establish an absolute relationship due to variations in "peculiar velocities" (velocities which are determined by gravitation influence). However, relationship between measured distance and measured redshift still suggested the possibility of an expanding universe.

[140B] *"Cepheid Variable Stars & Distance Determination;"* The Australia Telescope National Facility...
    http://www.atnf.csiro.au/outreach/education/senio/astrophysics/variable_cepheids.html
[140C] *"Shapley, Harlow "* Complete Dictionary of Scientific Biography, 2008, Charles Scribner's Sons.
    http://www.encyclopedia.com/topic/Harlow_Shapley.aspx

A problem occurred when Hubble's time of expansion, as determined by his data, proved incompatible with the earth's age as determined by the radioactive decay of rocks[141].

Hubble's empirically-determined value for his "constant" was $1.95556e9$ years. His theory had determined a time "constant" for which the universe was expanding of approximately 2 billion years. However, the half-life of radioactive decay in rocks which were encrusted in the earth measured four billion years since the radioactive material had been deposited.

After Hubble's death, the constant was shifted downward 90% so that "time since Big Bang" would be increased by an equivalent 90% :

$$H_O = 1/t; \quad \{ Let\ "H_O"\ become\ "70\ km/sec/Mpc \} \quad (70\ km/sec/Mpc) = 2.2687e-18 = 1/t;\ t = 13.97e9\ years$$

To reduce Hubble's Constant by 90%, increases "distance" by 90% and "time" by 90%, since time and distance must vary proportionally according to the Expanding Universe model.

If the Hubble constant, as determined by actual measurement, is incompatible with other-source time measurements, science should have recognized this fact as a defect in the expanding universe theory. However, since they had no alternative to an "expanding universe," they responded by deserting Hubble's empirical foundations for his constant in favor of a better time "fit."

The time-scheme shifting of Edwin Hubble's "constant" downward began only after his death in 1953. Hubble remained faithful to his measurements —to the data by which he had established his constant and recession time frame.

The 1929 Hubble Data Table Presumes Doppler Redshift and Estimates "H_O"

Object Name	Dist. (Mpc)	Vd. (km/s)	Object Name	Dist. (Mpc)	Vd. (km/s)	Object Name	Dist. (Mpc)	Vd. (km/s)
SMC	0.032	+170	5194	0.5	+270	1055	1.1	+450
LMC	0.034	+290	4449	0.63	+-200	7331	1.1	+500
6822	0.214	-130	4214	0.8	+300	4258	1.4	+500
598	0.263	-70	3031	0.9	-30	4151	1.7	+960
221	0.275	-185	3627	0.9	+650	4382	2.0	+500
224	0.275	-220	4826	0.9	+150	4472	2.0	+850
5357	0.45	+200	5236	0.9	+500	4486	2.0	+800
4736	0.5	+290	1068	1.0	+920	4649	2.0	+1090

SMC = Small Magellenic Cloud; LMC = Large Magellenic Cloud; All object numbers are preceded by "NGC." 1 parsec = 3.26 light years; 1 Mpc = megaparsec = 10^6 parsecs.

Edwin Hubble calculated his constant value by his 1929 data table[142] arraying redshift measurements from nearby galaxies to the distances which were measured by an

---

[141] Harvard web site. *op. cit.*
[142] *"A Relation between Distance and Radial Velocity among Extra-Galactic Nebulae"* by Edwin Hubble. 1929 PNAS Vol 15, Issue 3, pp. 168-173

independent method. Hubble primarily used individual Cepheid[143] stars he had detected in the foreign galaxies. Characteristically, Hubble data tables present redshift measurements as converted to "recession velocities." The formula for this conversion is absolutely quantum. Specifically, apparent recession velocity is the negation of the subdivision of the speed of light using the redshift value "Z" as the subdivisional unit:

$$v = \left(1 - \frac{1}{Z}\right) c \qquad \textit{as "Z" increases, "v" is a larger percentage of "c."}$$

The measurement of redshift is also characteristically quantum. The primary index is the Rydberg visible frequency (Balmer Series) absorption lines for hydrogen, hydrogen being the primary stellar material. The absorption lines for the Balmer Series are harmonically regular. In star light, the two highest frequencies in the Balmer Series (n'=8, n'=7) are output as light. The four lowest frequencies in the series (n'=6, n'=5, n'=4, n'=3) are "absorb light frequencies" (wavelengths missing in spectrum) :

Balmer Series Formula

$$\left(\frac{1}{2^2} - \frac{1}{n'^2}\right) \frac{1}{(\lambda_r = 91.14 \text{ nm})} = \frac{1}{\lambda}$$

Example

Calculated Wavelengths missing from spectrum	Measured Wavelengths (redshift=1.0036)
n'=6; 410.13 nm	n'=6; 411.61 nm
n'=5; 434.00 nm	n'=5; 435.56 nm
n'=4; 486.08 nm	n'=4; 487.83 nm
n'=3; 656.21 nm	n'=3; 658.57 nm

When Balmer Series missing wavelengths (absorption lines) are spectrographically measured at the wavelengths in the right column above, redshift "Z" is determined by dividing the measured wavelengths by the calculated wavelengths.

In his 1929 data table, Hubble had independent measures for distance to the galaxy (from Cepheid calculations) and redshift as measured above.

From these 24 independent determinations of redshift (given as velocity) and distance determined by Cepheids detected in the galaxies, Hubble selected approximately 14-17 cases to determine his constant value using the formula:

$$H_0 = \frac{v.}{d} \qquad \textit{v=Vd. and d=Dist. in table}$$

Hubble rejected data points which were obvious aberrations to the expansion pattern. For example, both the Large and Small Magellenic Clouds give much to high Constant values (H_0 =5312.5 for Small Magellenic Cloud ; H_0 =8529.4 for Large Magellenic Cloud) indicating possible high-energy event redshifts.

It is crucial to recognize the difference between continuous light redshift due to distance and high energy event redshift. Scientists are currently reporting redshifts in the range of "Z=6" with gamma-ray bursts and high-energy quasar events. A redshift of "6" would shift the highest visible light wavelength in Balmer series (388.9 nm) to the mid-infrared Brackett Series (2333.4 nm). All visible frequencies would be shifted out of the visible range into mid and far infrared. Before distance interpretations are made of such high-energy redshift events, scientist should explain why hydrogen-fusion bombs are also known to cause characteristic redshifts in the light flashes released[144].

In point of fact, redshift measurements of continuous-light sources are consistent with the

---

[143] Cepheids are pulsing, variable light-intensity stars. The period of pulse establishes a known and absolute brightness.

[144] Natural observation of the author.

quantum geometric maximum of "Z=1.571." NASA telescopes focused on continuos light sources measure such redshift ranges. The Sloan Digital Sky Survey (SDSS), is ongoing as of 2005 and aims to obtain measurements on around 100 million objects. The highest redshifts SDSS has recorded for galaxies (continuous light sources) is "Z=1.4.[145]" Further, the Two-Degree Field (2dF) Galaxy Redshift Survey of the Anglo-Australian Observatory measured redshift for 221 thousand Galaxies obtaining a maximum "Z" value of approximately 1.25. [146]

The Magellenic Cloud redshifts were much to high to explain by distance alone and were eliminated from the data.

Other data points were eliminated for a second reason. The galaxies NGC 6822, NGC 598, NGC 221, NGC 224, and NGC 3031 are not receding but closing towards us at velocities determined by blue shifting. For these galaxies, gravitational effect is producing an opposite motion to that of the proposed expansion. They also are rejected as aberrations to the hypothesis being tested.

The remaindered cases from the 1929 data table were used to estimate the constant. They are presented in the following table, with the redshift measures reinserted. . Nearly all redshift measurements were within the "threshold range" for redshift detection which the contemporary 2dF Galaxy Redshift Survey[147] had identified. That range was "lower case 'z[148]'< 0.003 and > 0.0000 (margin of error ±.0003)."

Presumed Recession Velocities

velocity $(km/s) = (1 - 1/Z)c$; *from Doppler equation*

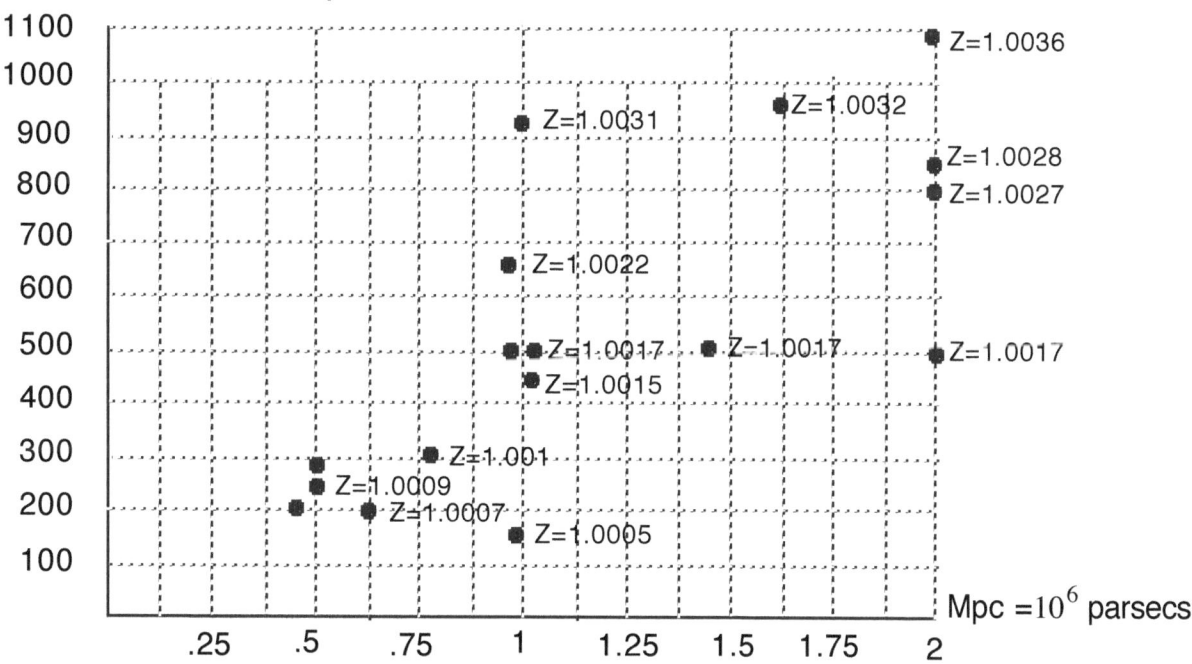

Source: *Hubble 1929 data table.*

[145] The Sloan Digital Sky Survey home page; *http://www.sdss.org/*
[146] The 2dF Galaxy Redshift Survey: Final Data Release, —June 30, 2003; *magnum.anu.edu.au/~TDFgg/*
[147] op. cit.
[148] lower case 'z'=Z-1 and represents the precentage of wavelength increase.

Hubble's conversion of redshift "Z" to apparent recession velocity is accurate, but not for the reason he supposed. It does not represent a real recession— as with his expanding universe concept— but is caused by light being forced to travel across the curvature of space. There is an "apparent recession velocity" because light travels over the arc appearing to slow down relative to the linear distance. From this data, an estimated value for the constant can be calculated for every data point. The value which Hubble settled upon "$H_o \approx 500$ km/ sec./ Mpc" is close to the the mean of the data points:

$\overline{x} = 472.2$ km/ sec./ Mpc

See data table below.

d (in Mpc)	v (in kg/s)	$H_o = \dfrac{v}{d}$	Var. $= H_o - \overline{H_o}$
2	1090	545	72.82
2	800	400	-72.18
2	850	425	-47.18
2	500	250	-222.18
1.7	960	565	92.82
1.4	500	357	-115.18
1.1	500	455	-17.18
1.1	450	409	-63.18
1	920	920	447.82
.9	500	556	83.82
.9	150	167	-305.18
.9	650	722	249.82
,8	300	375	-97.18
.63	200	317	-155.18
.5	270	540	67.82
.5	290	580	107.82
.45	200	444	-28.18
mean ($\overline{H_o}$)		472.17647059	
standard deviation $\sigma = \left(\dfrac{\sqrt{\sum \text{Var.}^2}}{\sqrt{n}} \right)$		177.11727587	

The current constant estimate is "65 km/sec/Mpc." It is 2.3σ (standard deviations) from Hubble's 1929 mean estimated constant of 472.2 km/sec./Mpc". Using the confidence interval for " 2.3σ (approx. .97)[149]," the current estimate of "H_0 =65" has only a 0.015 probability of occurring by chance within the Hubble data set[150] .

Further, the greatest redshift value for the Hubble data using the revisionist "65 km/sec/Mpc" constant value would only be Z=1.00043, almost outside the detectable range as determined by the 2dF Galaxy Redshift Survey. All other "Z" values from the Hubble data table, using "65 km/sec/Mpc" would fall outside the 2dF threshold.

After Hubble's death in 1953, assaults were made upon his measures, claiming that he had underestimated distances. Walter Baade argued that Hubble's Cepheids were part of "star clusters" and therefore had a much greater light intensity than originally estimated by Hubble. In the 1950's, Baade[151] had discovered the dimmer "Population II Cepheids." He had tried to recalibrate the "Classic Cepheids" which Hubble had used[152] — arguing Classic Cepheid's were brighter and Hubble's calculation of his constant too high. However, Baade's recalibration has not been universally accepted.

It wasn't until 1997 that an actual empirically founded challenge was made to the Cepheid brightness scale which Hubble had used. Feist and Catchpole took data from the parallaxes satellite, Hipparcos[153] and compared parallaxes triangulations to the nearest Classical Cepheids with distance measurements based upon the Cepheidic period. Their study proposed an upward revision of original Cepheid brightness by .2 magnitude[154] . The Feist and Catchpole revision would downshift Hubble's original calculations from a constant of "500 km/sec./Mpc" to nearer "400." However, the Feist and Catchpole revision does not fully meet the test of scientific reliability because the distances to nearby Cepheids (1000 light years or more) are on the verge of being outside parallaxes triangulation range for the Hipparcos satellite and therefore subject to error.

Curiously, it is only the Feist and Catchpole study which offers any experimental evidence to contradict Hubble's 1929 calculations. Yet that data was issued forty years *after* Hubble's Constant was revised downward by Sandage *et. al.* The Feist and Catchpole Cepheid brightness revision does not warrant anything like the 90% emendation which Hubble's Constant suffered after his death. Feist and Catchpole say their revision results in a 10% increase in distances. By extension, it would therefore result in a 10% decrease in Hubble's 1929 constant value.

---

[149] Weisstein, Eric W. "Standard Deviation." From MathWorld--A Wolfram Web Resource. http://mathworld.wolfram.com/StandardDeviation.html

[150] Below I will show that the revisionist constant of "65-70 km/s/Mpc" predicts distance measures for the 1929 Hubble data set which are so high as to be beyond the range of possibility.

[151] Wilhelm Heinrich Walter Baade (March 24, 1893–June 25, 1960) was a German astronomer who emigrated to the USA in 1931. He took advantage of wartime blackout conditions during World War II, which reduced light pollution at Mount Wilson Observatory, to resolve stars in the center of the Andromeda galaxy for the first time, which led him to define distinct "populations" for stars (Population I and Population II).

[152] Harvard University "The Hubble Constant" web page. Op. Cit.

[153] *"The Cepheid PL Zero-Point from Hipparcos Trigonometrical Parallaxes (1997);"* M. W. Feast, R. M. Catchpole; http://astro.estec.esa.nl/Hipparcos/pstex/feast_ceph.ps

[154] magnitude $= \left(100^{.2}\right)^n$ "standard candle"

The desertion of the rigorous empiricism of the 1929 data table for "age of the universe" schemes is misfortunate, if accuracy in measurement is the goal. Because quantum curvature measures redshift as apparent change in velocity, "age of the universe" is an irrelevant time factor, and Hubble's data can be evaluated independently.

Hubble's original data identifies a contradictory motion to apparent recession velocity. The "negative recessions" identified by blue shifting (NGC 6822, 598, 221, 224 and especially 3031) are motions of contraction due to gravitational influence. Most of the gravity "negative recessions" galaxies are the closest foreign galaxies, being in the ".2-.3 Mpc" distance range. There is one exception. NGC 3031 is at .9 Mpc, within the distance ranges used to estimate "H_o." Since gravitational force equals the multiple of the two masses divided by distance squared, "3031" is obviously of greater mass than the other closer "negative recession" galaxies.

Within the data set used to estimate the constant, a galaxy of sufficient mass can be affected by reverse gravitational motion and its "H_o" calculation will be low. Gravitational contraction provides a "negative variance bias" to the "H_o" estimations.

The greatest negative variation is NGC 4826 at a variance of -305.18 from the mean "$\overline{H_o}$." Significantly, "NGC 4826" is at the same distance (.9 Mpc) as "NGC 1068" which has a negative apparent recession velocity. This indicates that motion of galaxies at this distance can be influenced by gravitational interaction, depending upon their mass.

"Positive variation bias" to the "H_o" calculations is probably not explained by the Magellenic Cloud data which Hubble obviously included for that purpose. A better explanation is probably offered by the Baade observation that some of Hubble's Cepheid indicators may have been contained within star clusters and thus were brighter and gave an artificially low distance estimation. For example, NGC 1068 has a higher redshift measurement (higher apparent recession velocity) at distance of "1 Mpc" than all galaxies of greater apparent distance, except one.

This resulted in the highest positive variance from the mean for NGC 1068 at + 447.83 or two standard deviations from the experimental mean value of the constant. If the actual constant value were the mean, NGC 1068 would be "1.9484240688" times as far as the Cepheid measurement made it out to be. NGC 1068 would actually be four times as bright (indicating a cluster) than the single phase measured Cepheid it was assumed to be.

It is true that bias from Cepheid/ star-clusters misidentification will have greater influence on the mean than bias from gravitational negation of (apparent) recession velocity. The actual value of Hubble's Constant will be lower than the experimental mean.

However, the true constant could not possibly fall to the currently presumed value of "65 km/ sec./ Mpc." This is lower than the constant as calculated from gravitationally-biased "NGC 4826" (167 km/ sec./ Mpc). The true constant could not be lower than that calculated from a measured redshift — known to be biased by negative motion from gravity. Statistically, "NGC 4826" establishes the lowest possible limit of the true value of Hubble's Constant.

The post-Hubble revisions represent the desertion of empirical science in the defense of incorrect theory. The post-Hubble constant was not chosen by the empirical method by; by the measure of redshift determining (apparent) recession velocity and an independent measure of distance to the galaxy which originated that redshift:

$$H_o = \frac{(1 - 1/Z)\,c}{d}$$ *Accurate measures of "Z" and "d" give "H_o."*

Instead, the post-Hubble revision was chosen by an "age of the universe" time-factor required by the expanding universe model. That model is incorrect and the proof of this is that it has required abandonment of Hubble's 1929 data.

The quantum curvature model explains distance-proportional redshift and Hubble's mathematical description of it as an apparent difference in the velocity of the speed of light traveling a quantum-produced curvature over a linear distance.

Any distance, "d," between ourselves and stellar light source is a partial or subdivision of the quantum, "Q." "Q" establishes the diameter of the visible universe by providing maximum curvature which is the circumference of the semicircle with a "diameter=Q."

Any subdivision of "Q," designated as "d," Follows the quantum law of elliptical curvature. The distance "d" is kinked into an elliptical curvature with an eccentricity squared equal to the negation of the square of "d as subdivision of Q" [155] :

$$\frac{d}{Q} = \text{subdivision of Q} \; ; \; \varepsilon = \text{elliptical eccentricity}$$

$$\varepsilon^2 = 1 - \left(\frac{d}{Q}\right)^2$$ *The negation of subdivision^2 of Q*

This kinking of stellar distances into elliptical curvature is the actual source of the redshift as light follows the path of curvature. The redshift can be predicted for all values of "d" if the value of the quantum "Q" is known:

$$Z = \frac{\text{elliptical periphery}}{2d}$$

The peripheries of ellipses of known eccentricity and major axis can only be estimated using conventional three-dimensional Euclidean geometry[156] . However, an exact formula for the periphery of the ellipse has been developed using quantum geometry[157] :

$$\chi = \text{circumference of ellipse} \; ; \; r_1 = \text{minor axis} \; ; \; r_2 = \text{major axis}$$

$$\chi = 2\sqrt{3\,r_1^2 + r_2^2}\left(\frac{2\,r_1}{\sqrt{r_1^2 + 3\,r_2^2}}\left(\frac{\pi - 3}{3}\right) + 1\right) + 2\left(r_1\left(\frac{\pi - 3}{3}\right) + r_2\right)$$

This formula can be used to provide a predicted redshift "Z" for any stellar distance "d," as a subdivision of the diameter of the visible universe, "Q"[158] :

$$Z = \sqrt{\frac{3}{4}\left(\frac{d}{Q}\right)^2 + \frac{1}{4}}\left(\frac{2d}{\sqrt{d^2 + 3Q^2}}\left(\frac{\pi - 3}{3}\right) + 1\right) + \left(\frac{d}{2Q}\left(\frac{\pi - 3}{3}\right) + \frac{1}{2}\right)$$

Hubble's 1929 data table has proved inconsistent with the expanding universe model and was ignored by later generations of astronomers. They proposed a constant revision which, I will show, is completely incompatible with Hubble's 1929 data. That data confirms the

---

[155] *The Quantum Law of Ellipses and the Elliptical Kink* in Appendix

[156] The best estimation uses the MacLaurin Derivative Series which is based upon derivatives of eccentricity.

[157] *Quantum Determination of Elliptical Periphery and the Detection of Systemic Error in the Maclaurin Derivative Series;* master's thesis, The Virtual University; Dawson, Lawrence

[158] *The Quantum Formula for Redshift* p. 15

quantum curvature model and eliminates the post-Hubble revision.

Quantum geometry seeks a scientifically accurate measurement of the physical world. The 1929 data table represents the serious attempt of very competent astronomer to compare measured redshift with measured distances in the relatively close distances such measurements are possible. There is however a variance between measured distance and mathematically predicted distance for the measured redshift This is true whether that mathematical prediction is made by Hubble's expanding universe model and Doppler effect or it is made by the quantum curvature model. Clearly other factors are affecting the redshift measurement, the primary one being gravitational influence between nearby galaxies. Galaxies rushing towards one another from gravitational influence produce blue shifting of the light, a phenomenon clearly identified by Hubble's data (objects 6822, 598, 221, 224 and 3031 have antirecession velocities).

Any redshift measurement can be affected by extraneous influences and will generate a variance between measured distance and the distance mathematically predicted for the measured redshift. In the competition between the expansion universe mathematical model and the quantum curvature mathematical model, the model which best eliminates the variance is the correct one. Since variance is caused by modification of the predicted redshift at the distance, the more correct the mathematical determination, the less the mean variance will be. The quantum curvature model wins this competition.

Hubble's "expanding universe" constant and its predicted redshift at distance can be compared with redshift at distance predicted by quantum curvature.

Quantum Curvature Graph of Distance at Redshift (d=10 at Z=1.571)

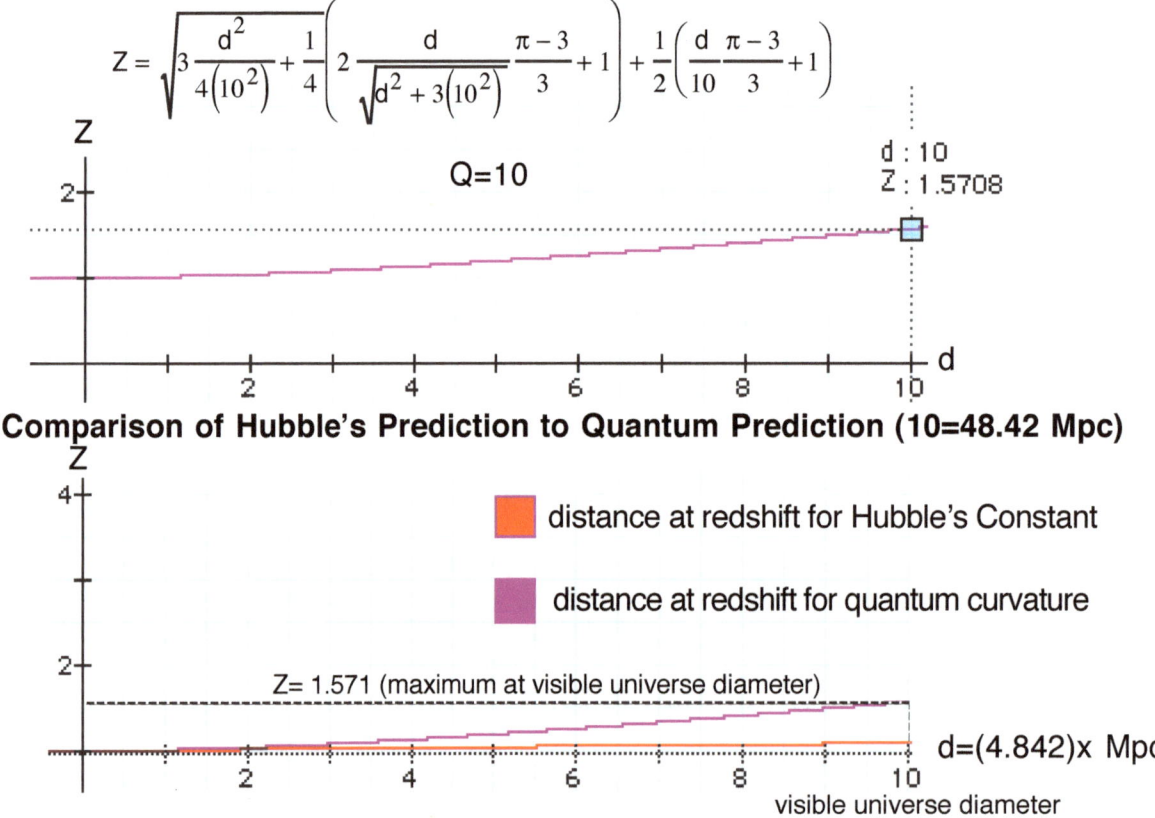

$$Z = \sqrt{3\frac{d^2}{4\left(10^2\right)} + \frac{1}{4}\left(2\frac{d}{\sqrt{d^2 + 3\left(10^2\right)}}\frac{\pi - 3}{3} + 1\right) + \frac{1}{2}\left(\frac{d}{10}\frac{\pi-3}{3} + 1\right)}$$

Comparison of Hubble's Prediction to Quantum Prediction (10=48.42 Mpc)

26

Hubble's 1929 data were used to find the most probable visible universe diameter for the quantum formula. Starting with Hubble's predicted distance at "Z=1.571" (217,8773 Mpc), trial-and-error was used to establish a quantum "Q" value which provided the best quantum curvature "fit" to the 1929 data. The best fit was found to be a partial of Hubble's predicted "Q" distance of 217.8773: Q=(217.8773)/ 4.5=48.42 Mpc

$$Q = \frac{1}{4.5}\text{Hubble at redshift Z=1.571.} \quad Q = \frac{1}{4.5}(217.8773) = 48.42 \text{ Mpc}$$

Hubble's Formulation; v=recession velocity ; H_o= Hubble's Constant

$$v = \left(1 - \frac{1}{Z}\right) c \qquad H_o = \frac{v.}{d} \; ; d = \frac{v}{H_o}$$

measured Z	predic. "d" Post-Hubble revision $H=70$ km/s/Mpc (in Mpc)	predic. "d" Hubble (in Mpc) $H=500$ km/s/Mpc	predic."d" elliptic. (in Mpc) $Q=48.42$ Mpc	Measured "d" (in Mpc) object name
1.0005	2.14	0.3	0.424	0.9 obj. *4826*
1.00057	2.42	0.34	0.4713	0.032 obj. SMC
1.00066	2.86	0.4	0.536	0.63 and 0.45 obj *4449/ 5357*
1.0009	3.86	0.54	0.707	0.5 obj *5194*
1.001	4.29	0.6	0.769	0.8 obj *4214*
1.0015	6.43	0.9	1.075	1.1 obj *1055*
1.0017	7.14	1	1.19	.9/ 1.1/ 1.4/ 2.0 obj *5236, 7331,4258, 4382*
1.0022	9.29	1.3	1.452	0.9 obj *3627*
1.0027	11.43	1.6	1.69	2.0 obj. *4486*
1.0028	12.14	1.7	1.74	2.0 obj *4472*
1.0031	13.14	1.84	1.878	1.0 obj *1068*
1.0032	13.71	1.92	1.92	1.7 obj *4151*
1.0036	15.57	2.18	2.09	2.0 obj. *4649*

The above table identifies several important facts. In the first place, the modern revised Hubble's constant of "65-70 km/s/ Mpc" is completely rejected by the empirical data. Alll distance prediction for "Z" values by the modernist revision are between 2 and 10 times the actual measured distances. The predicted distance for the lowest "Z" value (Z=1.0005) is greater than the furthest distance measured in the table. All variances are whole number multiples of measured distances. The modernist revision simply cannot estimate the measured distances with any credibility.

There is too much variance within the 1929 Hubble data set to establish anything but a trend. The higher redshift "Z" values tend to be at greater distances. However, the inclusion of the quantum curvature model along with Hubble's distance predictions may have refined that trend. Notice that the quantum predicted distances are higher than Hubble predicted distances starting at "Z=1.0005" and increasingly so until the two graphed lines reach "Z=1.0017." At this point, the two begin approaching one another again and cross at "Z=1.0032." After this point the Hubble predicted distances will always be greater than the quantum predicted distances. Tantalizingly, Hubble's predicted distances begin to pull away from measured distance at the "Z=1.0036." At higher redshifts from sources outside the measurable range for distances, Hubble's Constant may be giving unrealistically high distance predictions. The visible universe may actually be much smaller than modernists currently believe. If the quantum curvature model is correct, these data indicate that the edge of the visible universe may only be $1.5793 (10^8)$ light years away from the earth. This is in contrast to the $4.65 (10^{10})$ light years currently believed. Modernists may have overestimated the visible universe by a factor of 294 times.

Differences Between the Hubble and the Quantum Distance Predictions

measured Z	Variance squared (σ^2) between Hubble prediction and measured distance	Variance squared (σ^2) between quantum prediction and measured distance	Difference in predicted distance between Hubble and quantum $diff = d_Q - d_H$.
1.0005	0.36	0.227	0.124
1.00057	0.094864	0.1929	0.131
1.00066	0.0529 0.0025	0.008836 0.0074	0.136
1.0009	0.0016	0.042849	0.167
1.001	0.04	0.000961	0.169
1.0015	0.04	0.000625	0.175
1.0017	0.01 0.01 0.16 1	0.0841 0.0081 0.105625 0.6561	0.190
1.0022	0.16	0.305	0.152
1.0027	0.16	0.0961	0.09
1.0028	0.09	0.0961	0.04
1.0031	0.7056	0.770884	0.038
1.0032	0.0484	0.0484	0
1.0036	0.0324	0.0081	-0.09
$\overline{\sigma}^2$	**0.1746**	**0.1564**	

Expanding Universe May Be in Error

It is possibility that the expanding universe assumption, which has guided science for over 70 years, may be in error. This possibility is confirmed by the above table; by the comparison of the mean variance of Hubble's expansionist model and that of quantum curvature. The variance between Hubble's prediction and actual measured distance is greater than the variance between prediction and measurement for quantum curvature. The mean variance for Hubble's prediction is 12% greater, at 0.1746, than the mean variance for quantum curvature at 0.1564.

Quantum curvature is a better fit with Hubble's own data. This is not definitive proof in that the variance between the two means is not statistically significant using the two tailed t-test for matched pairs. However, the t-test may not be adequate under conditions which might be termed "biased variance." 60% of the variance between predicted and measured distance is on the low side. The measured distance is less than predicted and this bias favors the lower of the two tested models which, in this case, is the Hubble model. The t-test assumes that variance outside tested variables must be random and unbiased.

The possibility of a non-expanding universe is especially significant because the Hubble expansionist concept has proved detrimental to theoretical physics in general. Specifically, Einstein's "cosmological constant," which emerged from the field equations for General Relativity, had to be abandoned. The "cosmological constant" is a tension attached to space itself, sometimes described as "nonzero vacuum energy."

Einstein included the cosmological constant as a term in his field equations for general relativity because he was dissatisfied that otherwise his equations did not allow, apparently, for a static universe: gravity would cause a universe which was initially at dynamic equilibrium to contract. To counteract this possibility, Einstein added the cosmological constant. However, soon after Einstein developed his static theory, observations by Edwin Hubble indicated that the universe appears to be expanding......Since it no longer seemed to be needed, Einstein called it the "'biggest blunder" of his life, and abandoned the cosmological constant. However, the cosmological constant remained a subject of theoretical and empirical interest. Empirically, the onslaught of cosmological data in the past decades strongly suggests that our universe has a positive cosmological constant.[159]

Quantum geometry has identified the cosmological constant as a time force of separation sustaining the spatial quantum. Heaviside's magnetic permeability/ electric permittivity formula for free space equals time force ($1/c^2$) and is given as a field value, in Newtons per geometric unit of space. The constant force of time sustaining spatial vacuoles is shown to be the equivalent of .one Newton *per* meter of vacuum[160]. The "Newton" unit of force seamlessly integrates with the time force constant. This is not directly transferable to the constant as used by astronomers because their cosmological constant is not given in standard international units (SI units) for fields, in force per unit of area or volume.

---

[159] *Cosmological Constant;* Yale University Wiikipedia entry; en.wikipedia.org/wiki/Cosmological_constant#cite_note-Yale-0

[160] *The Quantum Dimensional Review of the Einstein vs Newton Gravitational Controversy;* Dawson, Lawrence, The Paradigm Company. ISBN 978-1516918096

ABOUT THE AUTHOR

Lawrence Dawson is an autodidact whose science department has become the internet. His formal academic training had ended in the political convulsions which deconstructed Columbia University in the late 1960s. As a student Fellow of the Faculty under the tutelage of Robert K. Merton, one of the founders of the sociology of science and perhaps the nation's best "meta-scientist," the author had watched the university recomposed by a radical epistemology which undermined the scientific method.

Prior to the 1960s student uprising, the linguistics of Ludwig Wittgenstein had begun to dominate the department. Wittgenstein taught that language could never test reality since all linguistic meaning was only a social consensus. Wittgensteinianism was compatible with an emerging scientific corruption which was replacing empiricism with a non-tested consensus. Only a few years previously, Hubble's constant, which had provided the foundation for the popular "Big Bang" theory and an expanding universe, had been revised downward by consensus even though the revision was incompatible with Hubble's original data set. The revision occurred because the "consensus" needed a longer age for the universe than Hubble's original constant had provided.

The author found himself intellectually paralyzed in the recomposed university which was replacing data and hypothesis testing with a Witttgensteinian generated social consensus as the means of determining scientific truth. That paralysis excluded him from an academic career.

It was 25 years later when, as an editor for a small academic publisher which was monitoring scientists who had lost employment due to unapproved research directions, that the author found the indisputable evidence for the damage that Wittgensteinian consensus had done to science. The 1995 Nobel Prize in chemistry had been given for a set of equations which had been disproved prior to the award. However, the disproving data had been completely suppressed in a consensus dominated scientific press and this suppression had allowed the Nobel to go forward. This discovery led to the book "The Death of Reality" which documented the damage which Wittgenstein had done to science and to the culture in general.

With that liberating discovery, the author felt free to pursue his original interest in scientific mathematics. This led to a critical review of the quantum mechanics which had been proposed in the early twentieth century. In turn, this review led to a rejection of the non-rational "consensus" mathematics which had supported primitive quantum mechanics. A rational system of quantum-dimensional mathematics based upon strict mathematical principles supplanted the consensus-based mathematics of primitive quantum mechanics, revised general relativity and eliminated big bang cosmology.

www.ingramcontent.com/pod-product-compliance
Lightning Source LLC
Chambersburg PA
CBHW050404180526
45159CB00005B/2147

* 9 7 8 1 5 1 9 2 3 7 1 3 2 *